All I want to Do is
Wear My Favorite
T-Shirt & Watch
Hallmark
Holiday Movies
With My Bff
All Day Long

My Notes

My Notes

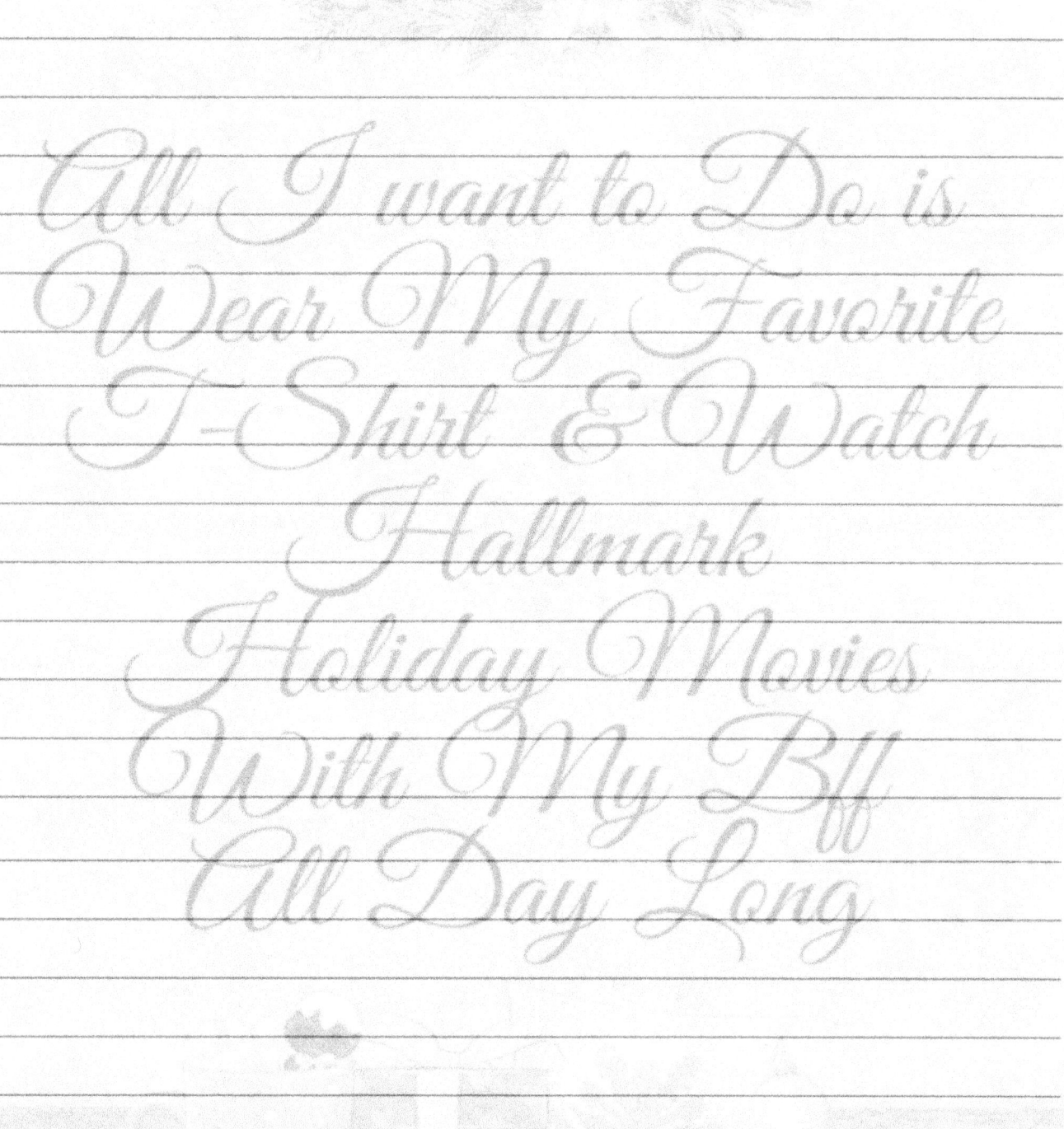

All I want to Do is
Wear My Favorite
T-Shirt & Watch
Hallmark
Holiday Movies
With My Bff
All Day Long

My Notes

All I want to Do is
Wear My Favorite
T-Shirt & Watch
Hallmark
Holiday Movies
With My Bff
All Day Long

My Notes

All I want to Do is
Wear My Favorite
T-Shirt & Watch
Hallmark
Holiday Movies
With My Bff
All Day Long

My Notes

All I want to Do is
Wear My Favorite
T-Shirt & Watch
Hallmark
Holiday Movies
With My BFF
All Day Long

My Notes

My Notes

All I want to Do is
Wear My Favorite
T-Shirt & Watch
Hallmark
Holiday Movies
With My BFF
All Day Long

My Notes

All I want to Do is
Wear My Favorite
T-Shirt & Watch
Hallmark
Holiday Movies
With My Bff
All Day Long

My Notes

All I want to Do is
Wear My Favorite
T-Shirt & Watch
Hallmark
Holiday Movies
With My Bff
All Day Long

My Notes

All I want to Do is
Wear My Favorite
T-Shirt & Watch
Hallmark
Holiday Movies
With My Bff
All Day Long

My Notes

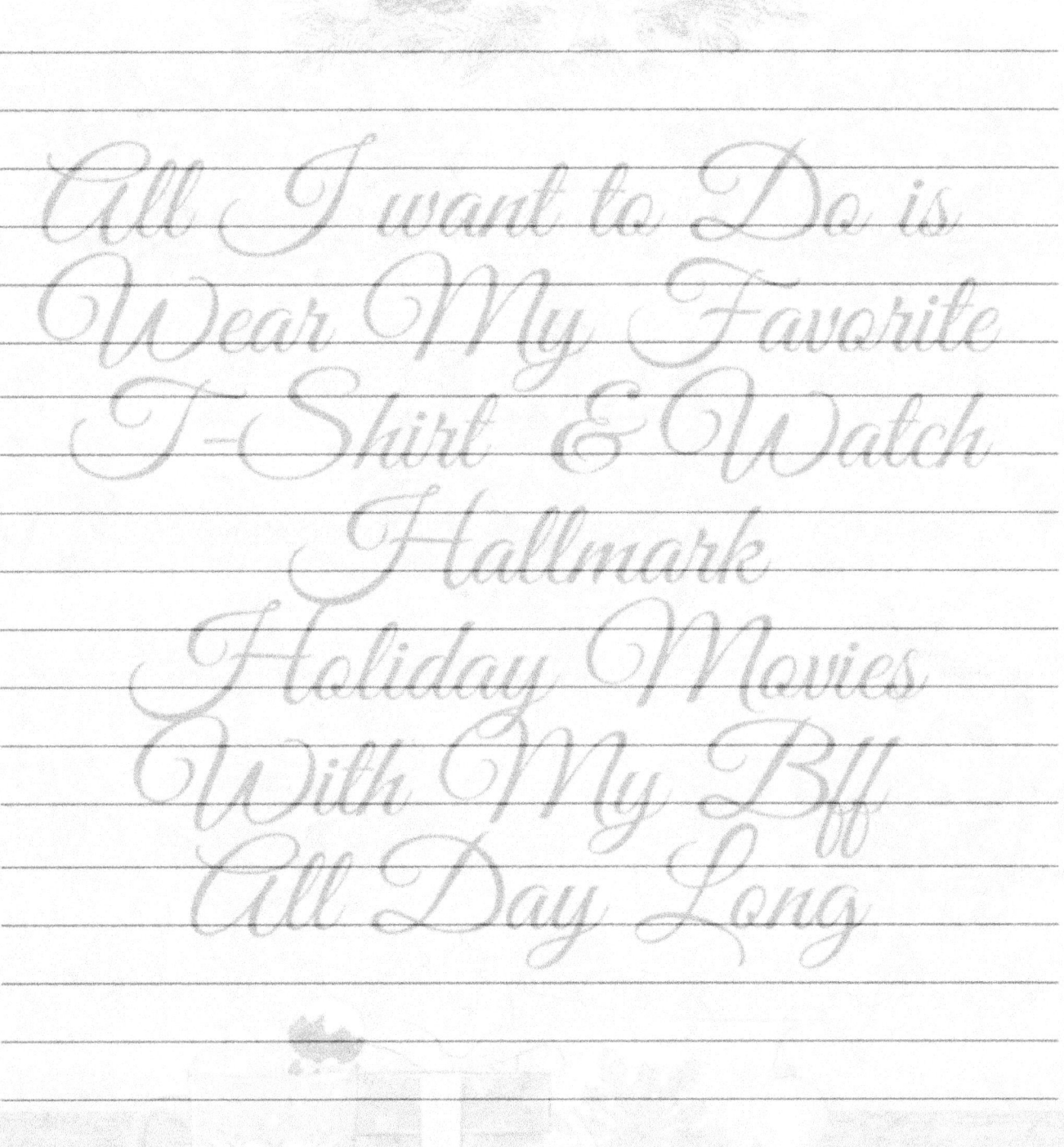

All I want to Do is
Wear My Favorite
T-Shirt & Watch
Hallmark
Holiday Movies
With My Bff
All Day Long

My Notes

My Notes

All I want to Do is
Wear My Favorite
T-Shirt & Watch
Hallmark
Holiday Movies
With My Bff
All Day Long

My Notes

All I want to Do is
Wear My Favorite
T-Shirt & Watch
Hallmark
Holiday Movies
With My Bff
All Day Long

My Notes

My Notes

All I want to Do is
Wear My Favorite
T-Shirt & Watch
Hallmark
Holiday Movies
With My Bff
All Day Long

My Notes

All I want to Do is
Wear My Favorite
T-Shirt & Watch
Hallmark
Holiday Movies
With My Bff
All Day Long

My Notes

My Notes

All I want to Do is
Wear My Favorite
T-Shirt & Watch
Hallmark
Holiday Movies
With My Bff
All Day Long

My Notes

All I want to Do is
Wear My Favorite
T-Shirt & Watch
Hallmark
Holiday Movies
With My BFF
All Day Long

My Notes

All I want to Do is
Wear My Favorite
T-Shirt & Watch
Hallmark
Holiday Movies
With My Bff
All Day Long

My Notes

All I want to Do is
Wear My Favorite
T-Shirt & Watch
Hallmark
Holiday Movies
With My Bff
All Day Long

My Notes

All I want to Do is
Wear My Favorite
T-Shirt & Watch
Hallmark
Holiday Movies
With My Bff
All Day Long

My Notes

All I want to Do is
Wear My Favorite
T-Shirt & Watch
Hallmark
Holiday Movies
With My Bff
All Day Long

My Notes

All I want to Do is
Wear My Favorite
T-Shirt & Watch
Hallmark
Holiday Movies
With My Bff
All Day Long

My Notes

All I want to Do is
Wear My Favorite
T-Shirt & Watch
Hallmark
Holiday Movies
With My Bff
All Day Long

My Notes

All I want to Do is
Wear My Favorite
T-Shirt & Watch
Hallmark
Holiday Movies
With My Bff
All Day Long

My Notes

All I want to Do is
Wear My Favorite
T-Shirt & Watch
Hallmark
Holiday Movies
With My Bff
All Day Long

My Notes

All I want to Do is Wear My Favorite T-Shirt & Watch Hallmark Holiday Movies With My Bff All Day Long

My Notes

All I want to Do is
Wear My Favorite
T-Shirt & Watch
Hallmark
Holiday Movies
With My Bff
All Day Long

My Notes

All I want to Do is
Wear My Favorite
T-Shirt & Watch
Hallmark
Holiday Movies
With My Bff
All Day Long

My Notes

All I want to Do is
Wear My Favorite
T-Shirt & Watch
Hallmark
Holiday Movies
With My Bff
All Day Long

My Notes

All I want to Do is
Wear My Favorite
T-Shirt & Watch
Hallmark
Holiday Movies
With My Bff
All Day Long

My Notes

All I want to Do is
Wear My Favorite
T-Shirt & Watch
Hallmark
Holiday Movies
With My Bff
All Day Long

My Notes

All I want to Do is
Wear My Favorite
T-Shirt & Watch
Hallmark
Holiday Movies
With My BFF
All Day Long

My Notes

All I want to Do is
Wear My Favorite
T-Shirt & Watch
Hallmark
Holiday Movies
With My Bff
All Day Long

My Notes

All I want to Do is
Wear My Favorite
T-Shirt & Watch
Hallmark
Holiday Movies
With My Bff
All Day Long

My Notes

All I want to Do is
Wear My Favorite
T-Shirt & Watch
Hallmark
Holiday Movies
With My Bff
All Day Long

My Notes

All I want to Do is
Wear My Favorite
T-Shirt & Watch
Hallmark
Holiday Movies
With My Bff
All Day Long

My Notes

All I want to Do is
Wear My Favorite
T-Shirt & Watch
Hallmark
Holiday Movies
With My Bff
All Day Long

My Notes

All I want to Do is
Wear My Favorite
T-Shirt & Watch
Hallmark
Holiday Movies
With My Bff
All Day Long

My Notes

My Notes

All I want to Do is
Wear My Favorite
T-Shirt & Watch
Hallmark
Holiday Movies
With My Bff
All Day Long

My Notes

My Notes

All I want to Do is
Wear My Favorite
T-Shirt & Watch
Hallmark
Holiday Movies
With My Bff
All Day Long

My Notes

All I want to Do is
Wear My Favorite
T-Shirt & Watch
Hallmark
Holiday Movies
With My Bff
All Day Long

My Notes

My Notes

All I want to Do is
Wear My Favorite
T-Shirt & Watch
Hallmark
Holiday Movies
With My Bff
All Day Long

My Notes

All I want to Do is
Wear My Favorite
T-Shirt & Watch
Hallmark
Holiday Movies
With My Bff
All Day Long

My Notes

All I want to Do is
Wear My Favorite
T-Shirt & Watch
Hallmark
Holiday Movies
With My Bff
All Day Long

My Notes

All I want to Do is
Wear My Favorite
T-Shirt & Watch
Hallmark
Holiday Movies
With My BFF
All Day Long

My Notes

All I want to Do is
Wear My Favorite
T-Shirt & Watch
Hallmark
Holiday Movies
With My Bff
All Day Long

My Notes

All I want to Do is
Wear My Favorite
T-Shirt & Watch
Hallmark
Holiday Movies
With My Bff
All Day Long

My Notes

All I want to Do is
Wear My Favorite
T-Shirt & Watch
Hallmark
Holiday Movies
With My BFF
All Day Long

My Notes

All I want to Do is
Wear My Favorite
T-Shirt & Watch
Hallmark
Holiday Movies
With My Bff
All Day Long

My Notes

All I want to Do is
Wear My Favorite
T-Shirt & Watch
Hallmark
Holiday Movies
With My Bff
All Day Long

My Notes

All I want to Do is
Wear My Favorite
T-Shirt & Watch
Hallmark
Holiday Movies
With My Bff
All Day Long

My Notes

All I want to Do is
Wear My Favorite
T-Shirt & Watch
Hallmark
Holiday Movies
With My Bff
All Day Long

My Notes

My Notes

All I want to Do is
Wear My Favorite
T-Shirt & Watch
Hallmark
Holiday Movies
With My Bff
All Day Long

My Notes

All I want to Do is
Wear My Favorite
T-Shirt & Watch
Hallmark
Holiday Movies
With My Bff
All Day Long

My Notes

All I want to Do is
Wear My Favorite
T-Shirt & Watch
Hallmark
Holiday Movies
With My Bff
All Day Long

My Notes

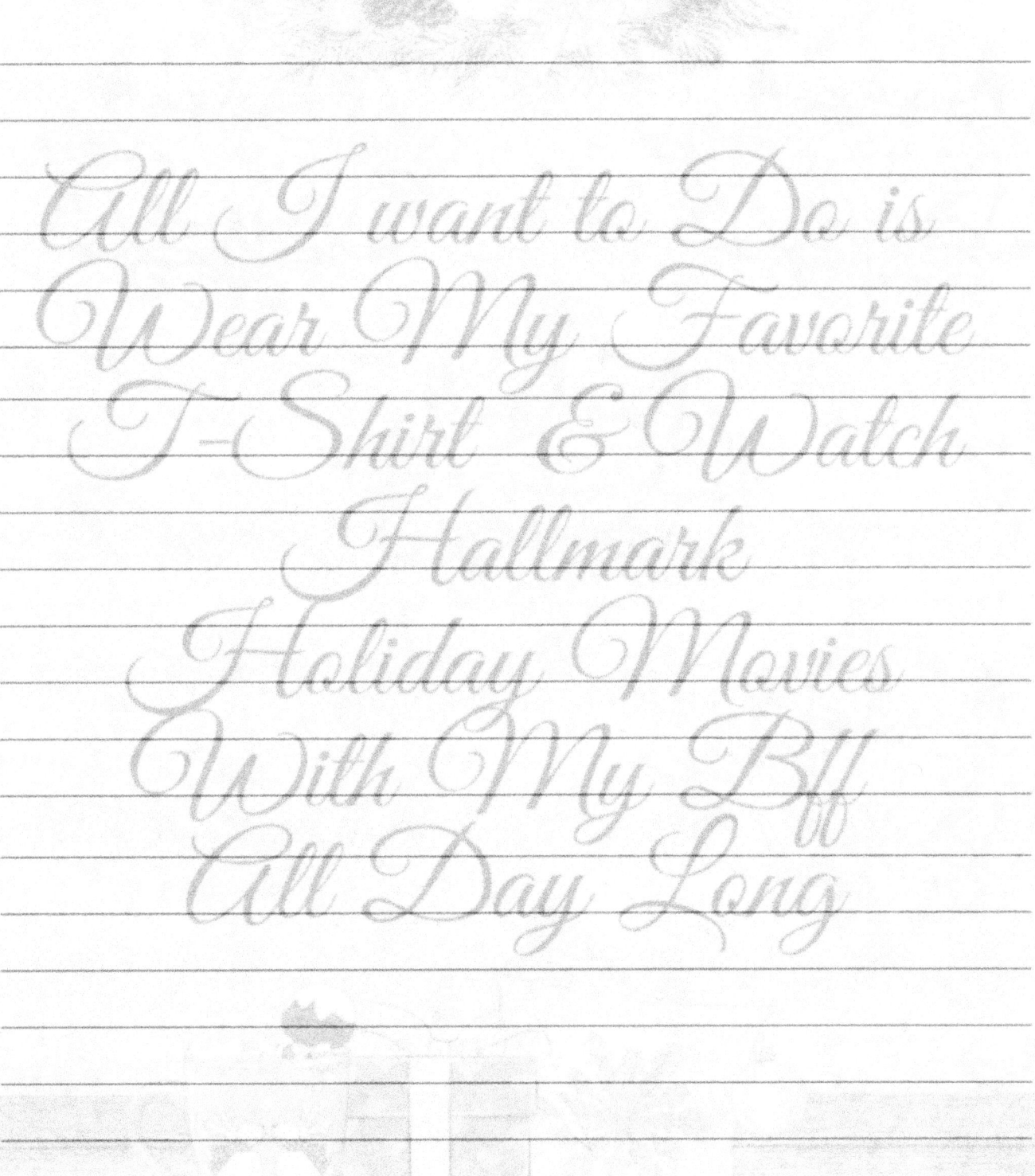

All I want to Do is
Wear My Favorite
T-Shirt & Watch
Hallmark
Holiday Movies
With My Bff
All Day Long

My Notes

All I want to Do is
Wear My Favorite
T-Shirt & Watch
Hallmark
Holiday Movies
With My Bff
All Day Long

My Notes

All I want to Do is
Wear My Favorite
T-Shirt & Watch
Hallmark
Holiday Movies
With My Bff
All Day Long

My Notes

All I want to Do is
Wear My Favorite
T-Shirt & Watch
Hallmark
Holiday Movies
With My Bff
All Day Long

My Notes

All I want to Do is
Wear My Favorite
T-Shirt & Watch
Hallmark
Holiday Movies
With My Bff
All Day Long

My Notes

_____

_____

_____

_____

_____

_____

_____

_____

_____

_____

_____

_____

_____

_____

_____

_____

_____

_____

_____

_____

_____

_____

All I want to Do is
Wear My Favorite
T-Shirt & Watch
Hallmark
Holiday Movies
With My Bff
All Day Long

My Notes